YOUR KNOWLEDGE HAS VALUE

- We will publish your bachelor's and
  master's thesis, essays and papers

- Your own eBook and book -
  sold worldwide in all relevant shops

- Earn money with each sale

Upload your text at www.GRIN.com
and publish for free

Suregada Multiflora Leaves Extract as a Green Corrosion Inhibitor

Abhik Chatterjee
Jiban Saha
Suvankar Debbarman
Dilliram Pokhrel

Bibliographic information published by the German National Library:

The German National Library lists this publication in the National Bibliography; detailed bibliographic data are available on the Internet at http://dnb.dnb.de.

ISBN: 9783346616128
This book is also available as an ebook.

STUDY OF SUREGADA MULTIFLORA LEAVES EXTRACT AS A GREEN CORROSION INHIBITOR ON IRON 1N HCl MEDIUM

Jiban Saha, Suvankar Debbarman, Dilliram Pokhrel, Abhik Chatterjee[*]

Corrosion Laboratory, Department of Chemistry, Raiganj University, Uttar Dinajpur,

ABSTRACT

Suregada multiflora leaves extract (SMLE) was prepared and used as a new eco-friendly green inhibitor for acid induced corrosion of mild steel(MS) in 1M HCl solution. The aqueous extract of leaves of *Suregada multiflora* was tested by dipping MS coupons in 1M HCl with and without SMLE solution at 303 K, 313 K, 333K and 343K using gravimetric method. The data obtained by weight loss measurement revealed that the SMLE has good inhibiton effect and mitigates the rate of corrosion. The inhibition efficiency (IE) increases with an increase in inhibitor concentration and with the exposure time. The adsorption study showed that the use of SMLE obeyed the Langmuir isotherm with regression co-efficient value nearly equal to unity. A part from this, Flory-Huggins isotherm and Langmuir-Freundlich isotherm were also used to study the interaction between the inhibitors and the mild steel surface, and mechanism of electrochemical reaction. At last the free energy of adsorption and activation energy were calculated and discussed.

Keywords: Corrosion inhibitor, Inhibition efficiency, Weight loss, Isotherm

Table of contents

1. **INTRODUCTION**

Metallic corrosion is one of the greatest problems in developed and developing countries due to unusual wastages of metallic instruments in the industrial sector [1]. Metallic corrosion arises due to environmental effects. Corrosion is unfavourable as many direct and indirect costs arise due to the damages such as productivity losses, interruptions, breakdowns, environmental pollutions, and even some legal actions [2].

About 2.5 trillion dollars which is the approximately 2.4% total gross of domestic products can be saved by corrosion inhibition of metals according to a worldwide study of the National Association of Corrosion Engineers (NACE) [3]. Among the metals, iron is one of the central metals and it alone contributes 60% to the industrial revolution. Iron and steel have been used almost in every field like transportation, construction, and machinery, etc [4].But the only problem is found for iron that it rusts easily and the corrosion products of iron may lead to construction collapse, leakage of fluids can lead to a serious accident and a threat to the environment. Especially in industry, acid is used to wash different machinery parts and in this acid pickling process, the machinery parts get deteriorated [5]. Nowadays acid pickling process becomes a major problem for expensive machinery parts. With proper corrosion prevention technologies, about 30% of losses in industries can be avoided. One of the expensive problems is the corrosion of the above metals in acid solution. In industry, hydrochloric acid has been used in metallic materials for chemical cleaning, de-scaling, pickling, etc [6].An acid environment promotes corrosion and it can't be prevented. Modern science can delay the corrosion process. Several methods can be used to mitigate corrosion. The best method to reduce corrosion is the use of inhibitors. A corrosion inhibitor represents the chemical compound that decreases the rate of corrosion of metallic substances when added to a gas or liquid phase. The inhibitors are mixed with solutions that are in direct contact with metal which prevents the anodic or cathodic reactions in an electrochemical cell that mitigates the corrosion [7]. The plant extracts are biologically acceptable, eco-friendly, or green corrosion inhibitors which can respond in a proper way to mitigate the corrosion without damaging the eco-system of the environment [8, 9]. In the market, large numbers of inhibitors are available. Different organic and inorganic compounds are used for these purposes. But few of them are environmentally friendly. Appropriate inhibitor selection is a very difficult task. Presently use of natural products like plant extracts is an emerging concept to make an eco-friendly planet. There are numerous studies where Plant-based corrosion inhibitors have been used [10, 11]. Recently green inhibitors attract the spotlight due to some of their advantages like they are eco-friendly, biodegradable, and nontoxic and does not contain any heavy metals [12-14]. Plant extracts are found to be rich sources of phytochemicals which can be a good substitution for traditional toxic inhibitors [15-17]. Leaves extract has been in the centre of interest of researchers due to its high content of phytochemicals compare to the other parts of the plants [18, 19]. In this work, Suregada multiflora leaves extract (SMLE) was studied as an environmentally friendly green corrosion inhibitor on iron metal in HCl solution by weight loss measurement.

2. EXPERIMENTAL

2.1 Material Preparation

The surface of rectangular pieces of iron with length 6.5 cm and width 2.5 cm were polished with emery paper and boiled for 15 minutes. The pieces were cleansed with double distilled water, dried in the air, and then in the oven. The weight of dry pieces was taken. The same process was repeated until getting constant weight.

2.2 Preparation of Leaves Extract

The leaves of *Suregada multiflora* plant were collected from the Botanical garden of Raiganj University. The leaves were washed with distilled water and are dried in an oven. Then about 15.0 g of leaves were weighed and boiled in 300 ml double distilled water to reduce the volume to 150ml. Again 75 ml distilled water was added and reduced to 200 ml. Finally, it was made cool and it was filtered with Whatman 40 filter paper, and preserved for the study.

2.3 Weight Loss Study

1M HCl solution was prepared by mixing the acid in distilled water and the corrosion study was carried out using 90 ml of the suitable corrosion medium. Initially, the weight of the specimen was measured and the required quantity of inhibitor was added to a 100ml glass beaker containing 90ml of synthetic corrosive medium and iron coupon. Experiments were conducted at 303 K, 313 K, 333 K, and 343 K temperatures. The time of contact during the study for a particular concentration of inhibitor was varied from 1 to 24 hours (i.e. 1, 2, 4,8,16 and 24 hours). Similarly, the experiment was repeated for the same period by altering the concentration of the inhibitor. The weight loss of the specimen was measured using a digital balance (K- Roy). The efficiency of the inhibitor was then expressed by using the data of the experiment.

The inhibition efficiency ($\eta\%$) measured by the following equation

$$\eta\% = \frac{\Delta W_{ui} - \Delta W_i}{\Delta W_{ui}} \text{ X } 100 \tag{1}$$

Surface coverage (θ) measured by the following equation-

$$\theta = \frac{\Delta W_{ui} - \Delta W_i}{\Delta W_{ui}} \tag{2}$$

Corrosion rate (*CR*) is measured by the following equation

$$\text{Corrosion rate}(CR) = \frac{W_0 - W}{A \times t} \tag{3}$$

Where ΔW_i *and* ΔW_{ui} are lost in the weight of the metal with and without inhibitor respectively. Similarly, W_0 and W are the initial and final weight of the iron coupon respectively, and A is the area of the iron coupon, and t stands for immersion time [20].

3. RESULTS AND DISCUSSION

3.1 Weight loss measurement

Weight loss experiments show that by changing the concentration of corrosion inhibitors both the rate and inhibition efficiency change. The η %, CR, and surface coverage (θ) can be calculated by using a gravimetric method that helps to determine the effectiveness of corrosion inhibition.

3.1.1 Exposure time and inhibition efficiency

Exposure or immersion time studies were performed to explore the stability of the corrosion inhibitor film as well as the rate of adsorption. The inhibition efficiency with various exposure times of iron in presence of a fixed concentration of inhibitors is presented in figure 1. The inhibition efficiency is determined from the weight loss measurement of mild steel in 1M HCl for 0.55g inhibitor concentration per 100 ml of corrosive medium and at 30°C temperature. It is clear from Figure-1 that the inhibition efficiency of *Suregada multiflora* on iron coupon raises with an increase in exposure time in the HCl medium. The main cause behind this is the formation of a very thin coating layer of inhibitor molecules on the metal surface.

There is a gradual rise in inhibition efficiency with rising immersion time and after 16 hours it decreases. The maximum corrosion inhibition efficiency was attained as a result of rapid adsorption of the inhibitor on the mild steel surface due to the presence of a greater number of active inhibitor molecules. The inhibition efficiency was found to be minimum (79.3%) after one hour of exposure period and it became maximum (96.28%) at 16 hours of exposure time.

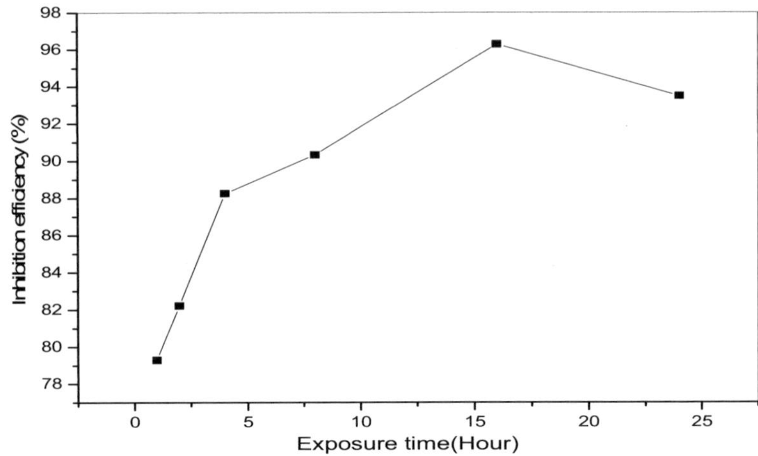

Fig. 1: Plot between Inhibition Efficiency with Exposure time

3.1.2 Inhibitor concentration and inhibition efficiency

The findings of gravimetric experiments with various inhibitors concentrations at different temperatures are given graphically as follows.

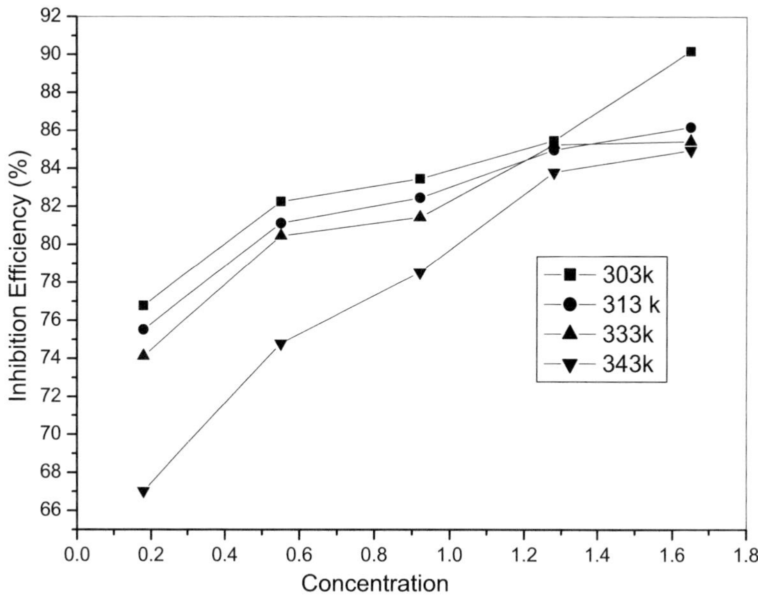

Fig. 2: A plot between Inhibition efficiency and Inhibitor concentration

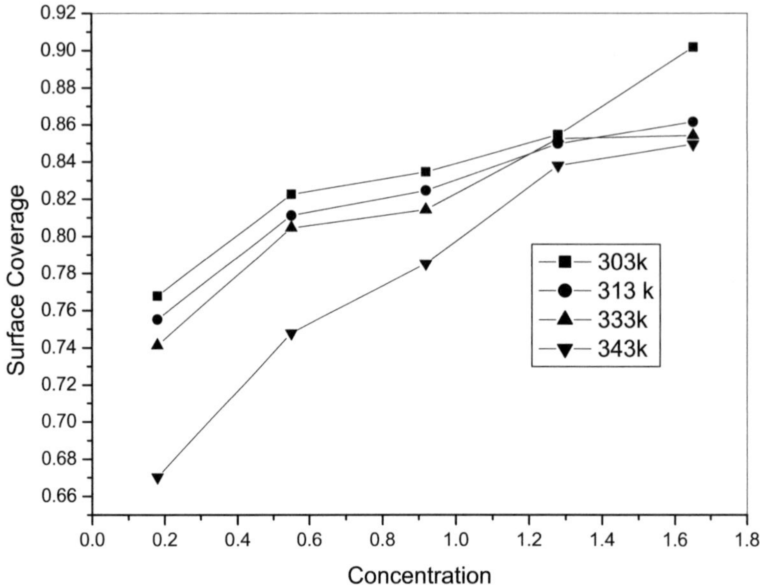

Fig. 3: A plot between Inhibitor Concentration and Surface Coverage

Figure 2 & 3 shows that the inhibition efficiency, as well as surface coverage raises with rising in the concentration of inhibitors in the HCl environment. It is due to the increase in the number of inhibitor molecules on the iron–acid solution interface.

3.1.3 Inhibitor concentration and corrosion rate

The corrosion-resistant value of *Suregada multiflora* was evaluated by adding the extract in 1.65, 1.28, 0.92, 0.55 and 0.18g/100ml concentration in corrosive medium (1M HCl). Figure 4 shows the plot between rates of corrosion with a concentration of the inhibitor. The experimental data revealed that the rate of corrosion decreases with a rise in concentration. The corrosion rate of the metal is considerably decreased from 0.00070 for low inhibitor concentration solution to 0.000278 at higher inhibitor concentration solution at 303K.Similar trend is also observed at high temperatures because of the gradual adsorption of inhibitor molecules on the metal surface and they form an organic barrier between the iron surface and the aggressive environment [20-22].

Table 1. Corrosion rate at different temperatures and different inhibitors concentrations

Inhibitor concentration (g/100ml)	Corrosion Rate (303K) (g cm^{-2}h^{-1})	Corrosion Rate (313K) (g cm^{-2}h^{-1})	Corrosion Rate (333K) (g cm^{-2}h^{-1})	Corrosion Rate (343K) (g cm^{-2}h^{-1})
1.65	0.000278	0.000809	0.00183	0.00214
1.28	0.000412	0.000891	0.00186	0.00231
0.92	0.00050	0.000921	0.00234	0.00306
0.55	0.00056	0.0011	0.00246	0.00359
0.18	0.00070	0.00151	0.00326	0.0047
0	0.00284	0.00771	0.01261	0.01425

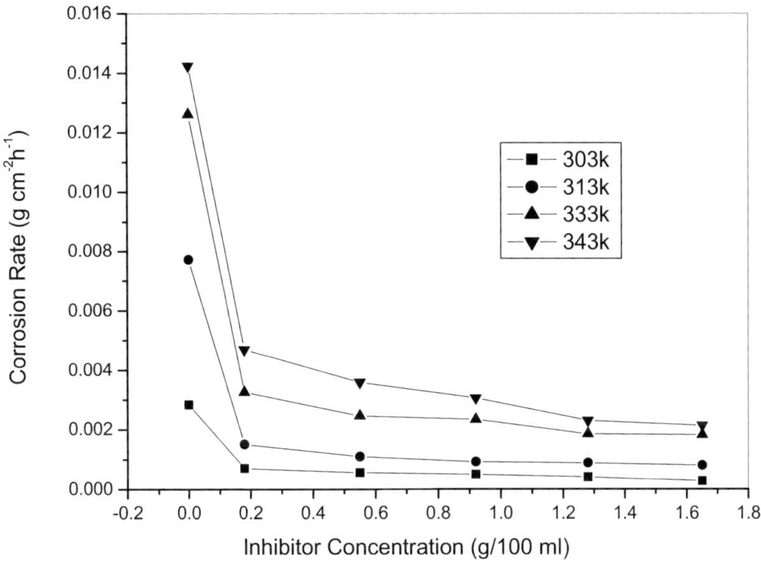

Fig. 4: A Plot between Corrosion rate and inhibitor concentration

3.2 Adsorption isotherms

The adsorption isotherm provides general information about the interaction between the inhibitors and the metal surface, and also the mechanism involved in it. There are different types of adsorption isotherm to explain the nature of interaction among the inhibitors and metal surface. Some adsorption isotherms are explained as follow:

3.2.1 Langmuir isotherm

The Langmuir isotherm relates the surface coverage (θ) and the inhibitor concentration (C_i) (g/100 ml), which are given as:

$$\theta = \frac{K_{ads}C_i}{1+K_{ads}C_i} \tag{4}$$

Where K_{ads} represents the adsorption coefficient and is given as:

$$K_{ads} = \frac{1}{55}e^{\frac{-\Delta G}{RT}} \tag{5}$$

a modified Langmuir adsorption isotherm is given by the following expression:

$$\pi = \frac{n}{K_{ads}} + nC_i \tag{6}$$

Where n is a constant. Figure 5 represents a plot between C_i/θ and C_i. The Langmuir adsorption isotherm shows a straight line with the regression coefficient (R^2) close to 1 for the media (i.e. 1M HCl). It represents the Langmuir isotherm that has been observed by the best fitting of the gravimetric results among the several adsorption isotherms. It also confirms that there is the formation of a monolayer of inhibitors on the surface of mild steel that resists corrosion.

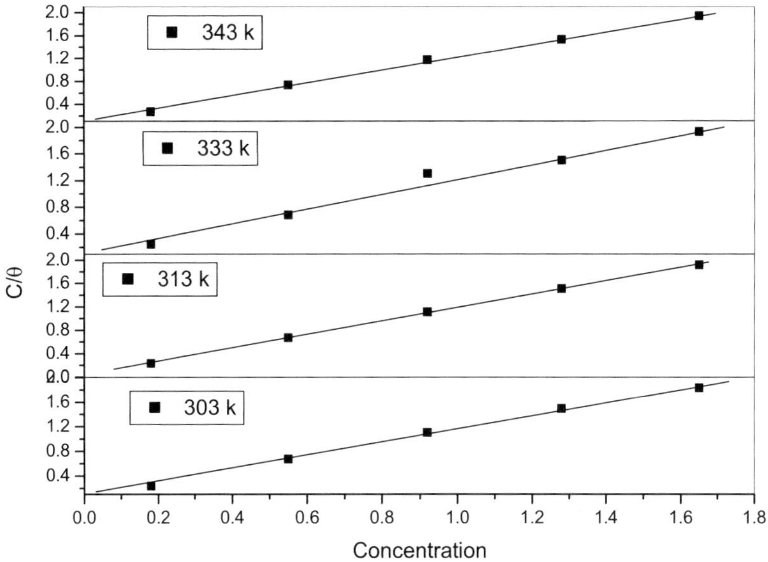

Fig. 5: Langmuir adsorption plots in 1 M HCl solution containing different concentrations of inhibitors at different temperatures

The value of linear regression coefficient (R^2) for the leaves extract at different temperatures of Langmuir Adsorption isotherm is given in Table 2:

Table 2. Regression coefficient values at different temperatures

Temperature	Regression coefficient (R^2)
303K	0.996
313K	0.999
333K	0.976
343K	0.997

3.2.2 Langmuir–Freundlich isotherm

The Langmuir–Freundlich isotherm (Figure 6) is provided on the basis of the following equation:

$$\left[\frac{\theta}{1-\theta}\right]^{\frac{1}{h}} = K_{ads}C$$

(7)

The heterogeneity parameter 'h' is a measure of the adsorption energy in different sites on the surface.

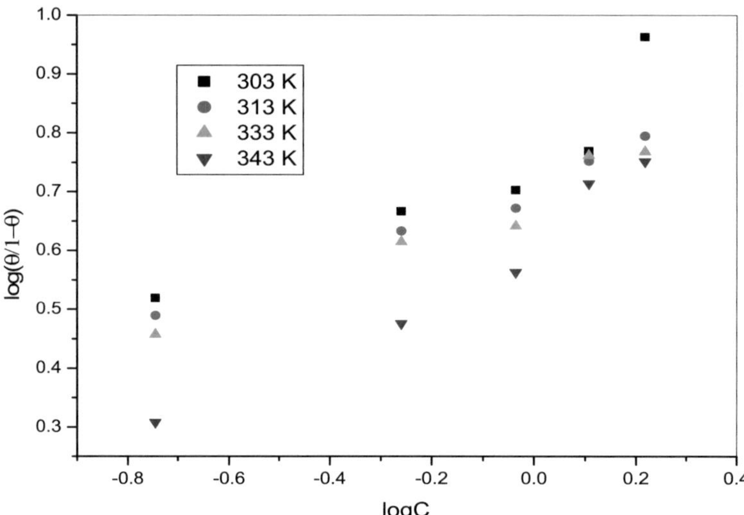

Fig.6: Langmuir–Freundlich isotherm for corrosion inhibition of mild steel in an acid environment.

3.3 Thermodynamics parameters

Adsorption of the inhibitor molecules is related to different factors like charge and the nature, the electronic characteristics of metal surface, adsorption of solvent, and other ionic species. The isotherm study helps to describe the adsorption mechanism of inhibitors on the metal surface. Commonly used adsorption isotherms are Langmuir, Flory-Huggins, and Freundlich-isotherms. Among them, Langmuir adsorption isotherm is useful to explain the best description of the adsorption behaviour (Fig.5). Hence the Langmuir adsorption isotherm is suitable to describe the adsorption of *Suregada multiflora* inhibitor on an iron coupon surface and from this isotherm, we can calculate the ΔG_{ads}. The Langmuir adsorption isotherm is given as:

$$\frac{C_i}{\theta} = \frac{n}{K_{ads}} + nC_i$$

(9)

Where C_i and K_{ads} are inhibitor concentration and equilibrium constant of adsorption respectively. C_i is related to the standard free energy of adsorption ΔG_{ads} as:

$$K_{ads} = \frac{1}{55.55}\exp\left(\frac{-\Delta G_{ads}}{RT}\right) \qquad (10)$$

Where R is the universal gas constant and T is the absolute temperature.

The free energy of adsorption (ΔG_{ads}) were determined by using the following equation [23]

$$\Delta G_{ads} = 2.303RT\log(55.55K) \qquad (11)$$

The data present in table 3 shows that the adsorption is mixed type because the value of ΔG_{ads} at different temperatures was found between -20 and -40 kj/mol.

Table 3. Values of free energy of adsorption (ΔG_{ads}) at different temperatures

Temperature	303K	313K	333K	343K
ΔG_{ads}(kj/mol)	-30.66	-26.55	-31.94	-32.89

The negative values of ΔG°_{ads} indicate the spontaneity of the inhibition process. If the ΔG_{ads} values are below -20 kJ mol^{-1}, it indicates the physisorption process whereas the chemisorptions process indicates when ΔG°_{ads} close to or higher than -40 kJ mol^{-1}. In this experiment the calculated values are -30.66, -26.55, -31.94, -32.89 respectively. So, this process represents the mixed type of adsorption.

3.4 The Activation energy (E_a)

Arrhenius equation to calculate the thermodynamic parameters of the corrosion reaction such as activation energy, and its alternative formulation called transition state equation is given as:

$$C_R = k \exp\left(\frac{-E_a}{RT}\right) \qquad (12)$$

Apparent activation energy values in the presence and absence of different inhibitor concentrations at different temperatures were calculated by linear regression between $\ln(C_R)$ and $(1/T)$ plots and listed in Table 4.

Table 4. Values of the calculated activation energies

Concentration (g/100ml)	1.65	1.28	0.92	0.55	0.18	Blank
E_a(kJ/mol)	40.477	34.809	37.892	37.54	38.083	30.903

The activation energy parameters showed that the value of Ea for inhibited solution is higher than that for uninhibited solution. The greater the value of Ea for inhibited solution represents that more energy barrier is required for the reaction to occur [24]. Generally the activation energy values are increased in the presence of inhibitors. It is proved that physisorption is involved in this experiment as the value of activation energy is inhibited system increases higher than in the blank solution.

4. CONCLUSION

The *Suregada multiflora* leaves extract (SMLE) has excellent inhibition properties for mild steel corrosion in 1M HCl solution. The inhibition is due to the film formation on the metal/acid solution interface as a result of adsorption of SMLE molecules. The inhibition efficiency increases with the increase in inhibitor concentration and decreases with increase in temperature. The highest efficiency was found to be 90.18% on 1.65 g/100ml at 300 K. The three adsorption isotherms namely Langmuir adsorption isotherm, Flory-Huggins isotherm and Langmuir-Freundlich isotherm were used to test the data. Among them Langmuir isotherm fits the data well with regression co-efficient (R^2) nearly equal to 1. The value of free energy of adsorption showed that the adsorption is mixed type. Thus aqueous extract of SMLE is considered to be a cheap, eco-friendly & effective corrosion inhibitor for mild steel in acidic media.

5. REFERENCES

1. Dehghani A, Bahlakeh G, Ramezanzadeh B, Ramezanzadeh M. *Construction and Building Materials*, 2020; **245**, 118464.
2. Ogunleye OO, Arinkoola AO, Eletta OA, Agbede OO, Osho YA, Morakinyo AF, Hamed JO. *Heliyon*, 2020; **6(1)**,3205-3217.
3. Goyal M , Kumar S , Bahadur I , Verma C , Ebenso EE. *J Mo Liq ,* 2018;**256**,565-573.
4. Santana CA, da Cunha JN, Rodrigues JGA, Greco-Duarte J, Freire DMG, D'Elia E. *J Braz Chem Soc*, 2020;**31(6)**, 1225–1238.
5. Zaher A, Chaouiki A, Salghi R, Boukhraz A, Bourkhiss B, Ouhssine M. *Int J Corros,2020*; 1-10.
6. Ahamed I, Quraishi MA. *Corros Sci* ,2009;**51**, 2006-2013.
7. Olanrewaju A, Oluseyi AK, Ayomide BV, Theresa AO, Oluwakayode AS. *IOP Conf Ser Mater Sci Eng,* 2019;***509*(1),** 1–9.
8. Bentiss F, Traisnel M, Vezin H, Hildebr HF, Lagrenee M. Corros Sci , 2004; **46**, 2781-2792.
9. Lagrenee M , Mernari B , Bouanis M , Traisne M, Bentiss M. *Cooros Sci,* 2002; **44(3),** 573-588.
10. Abdallah M , Altass HM, Jahdaly BAA.*Green Chem Lett Rev*, 2018;**11**,189-196.
11. Karthiga N, Rajendran S, Prabhakar P, Rathish RJ. *Int J nano corros sci eng.* 2015; **2(4)**,31-49.
12. Krishnaveni K, Ravichandran J, Selvaraj A. *Acta Metall Sin Engl Lett,* 2013; **26(3)**, 321-327.
13. Raja PB, Sethuraman MG. *Pigment Resin Technol,* 2009; **38**, 33-37.
14. Obot IB, Obi-Egbedi NO. *J Appl Electrochem*, 2010; **40(11),** 1977-1984.
15. Badiea AM, Mohona KN. *J Mater Eng Perform*, 2009; **18**, 1264-1271.
16. Li XH, Deng SD, Fu H *J Appl Electrochem,* 2010; **40(9),** 1641-1649.
17. Noor EA. *J Appl Electrochem*, 2009; **39(9)**,1465-1475.
18. Verma C, Ebenso EE, Bhadur I, Qurashi MA. *J Mol Liq*, 2018; **266**,577-590.
19. Buchweishaija J.*Tanz J Sci*, 2009;**35**,77-92.
20. Zaher A, Chaouiki A, Salghi R, Boukhraz A, Bourkhiss B, Ouhssine M. *Int J Corros*, 2020; 1-10.
21. Sehmi A, Ouici HB, Guendouzi A, Ferhat M, Benali O, Boudjellal F. *J Electrochem Soc,* 2020;**167(15)**, 155508.
22. Singh A, Talha M, Xu X, Sun Z, Lin Y. *Acs Omega*, 2017; **2(11)**, 8177–8186.
23. Benarioua M, Mihi A, Bouzeghaia N, Naoun M. *Egypt J Pet*, 2019;**28(2)**, 155–159.

24. Singh P, Ebenso EE, Olasunkanmi LO, Obot IB, Quraishi MA. J Phys Chem C, 2016; **120**, 3408–3419.